Elivaldo Nunes Modesto Junior
Aldejane Prado
Carmelita Ribeiro

Bioactive compounds: added to buffalo milk Greek yogurt

Elivaldo Nunes Modesto Junior
Aldejane Prado
Carmelita Ribeiro

Bioactive compounds: added to buffalo milk Greek yogurt

Preparation, physicochemical and microbiological characterization

ScienciaScripts

Imprint

Cover image: www.ingimage.com

This book is a translation from the original published under ISBN 978-613-9-68661-2.

Publisher:
Sciencia Scripts
is a trademark of
Dodo Books Indian Ocean Ltd. and OmniScriptum S.R.L publishing group

120 High Road, East Finchley, London, N2 9ED, United Kingdom
Str. Armeneasca 28/1, office 1, Chisinau MD-2012, Republic of Moldova, Europe
Printed at: see last page
ISBN: 978-620-8-22618-3

CONTENTS

PREFACE

The production of yogurt is an alternative for the use of buffalo milk, offering a food with high nutritional and functional value (CUNHA NETO et al., 2005). Buffalo milk is a food with high nutritional potential and some of the nutrients, when subjected to fermentation, undergo biochemical changes, increasing the digestibility and absorption of proteins, lipids and carbohydrates necessary for human metabolism, as well as being an excellent source of calories, especially for children.

Borges, Medeiros and Correia (2009) showed that buffalo milk yogurt has a firm texture, without needing to be strengthened with defatted dry extract (DDE), which means that a derivative with a higher economic yield can be obtained, with repercussions on consistency and taste. This can be attributed to the higher fat and protein content of buffalo milk compared to bovine milk.

The Marajó archipelago has a great natural diversity in general, and its potential for producing fruit of the most varied characteristics stands out. Specifically in the municipality of Salvaterra, Marajó Archipelago, Para, some villages stand out for their great variety of fruit production, which often spoils due to a lack of help with storage techniques and a lack of knowledge of these processes on the part of the producing families. According to Lima, Melo and Lima, (2002) fruit consumption is associated with a reduction in the incidence of cancer, lower blood pressure and cardiovascular events and this association has been attributed mainly to its antioxidant content.

Some fruits and vegetables are considered functional foods because they contain substances that benefit health, such as phenolic compounds (flavonoids, tannins, among others) and vitamin C, which are considered antioxidant compounds (BROINIZI et al., 2007, AJAIKUMAR et al., 2005). These substances are widely distributed in the plant kingdom, for example in the pitanga fruit, which in the Marajó Archipelago is widely known as sour cherry.

The sour cherry (*Eugenia uniflora* L. - Myrtaceae) is a plant species endemic to the Norte Fluminense region (ASSUMPÇÂO and NASCIMENTO, 2000) and has phenolic substances with antioxidant, hypoglycemic and antirheumatic actions. It is also used for stomach disorders and as an antihypertensive (AURICCHIO; BACCHI, 2003). Brazil is the place where the plant grows best, reaching up to eight meters in height (OLIVEIRA, 2006). In the North, the climate is very conducive to its cultivation and in Salvaterra the production of sour cherry, as it is known in the region, is not very widespread, being found mainly in the backyards of local residents and even on public roads.

According to Oliveira (2006), the Eugenia genus is among the most important in the Myrtaceae family, with species of commercial, nutritional and medicinal value. In the northeast of Brazil, its economic importance is growing due to its excellent sensory properties and the high levels of vitamin A and C in the fruit (210 mg/100g and 14 mg/100g, respectively).

CHAPTER I

BUBALINOCULTURE

General aspects

Bubalinoculture is an activity that has been gaining prominence in Brazil due to its capacity for animal labor, as well as meat and milk production (SANTINI et al., 2013).

According to MAPA (2016), the Brazilian buffalo herd is estimated at around 1.15 million, with the North being the country's largest producer with 720,000 animals, followed by the Northeast with 135,000 head and then the Southeast with 104,000 head. The Brazilian Association of Buffalo Breeders (ABCB, 2016) estimates a herd of three million head of buffalo in Brazil. This discrepancy between data is often due, according to Bernardes (2006), to the fact that the data is underestimated, because the registration of buffaloes is confused with that of cattle during registration.

The Marajó archipelago has the highest concentration of buffalo in Brazil. According to the IBGE's Municipal Livestock Survey (PPM), Pará had around 520,000 head (38% of the national total) in 2016, of which more than 320,000 were on the north and northeast coast of the island. The municipality of Soure had around 74,500 head of buffalo in 2017.

Although the northern region is the largest producer of buffalo in Brazil, its production is geared towards the meat market, unlike the southeastern region, which focuses on milk production (ANDRIGHETTO, 2011).

Definition of milk

According to Normative Instruction 62/2011, without further specification, milk is the product of complete and uninterrupted milking, under hygienic conditions, from healthy, well-fed and rested dairy cows. Milk from other animals must be called according to the species from which it comes (BRASIL, 2011).

Buffalo milk

The main characteristics of buffalo milk are its high protein content (25.55%), which is higher in essential amino acids than cow's milk (VERRUMA; SALGADO, 1994), its high fat content, which is higher in calories than bovine milk, and its high percentage of minerals, the most important of which is calcium. Bubaline milk has undesirable sensory characteristics when consumed in its entirety. This situation is reversed when fat is standardized to between 2% and 3% (BUZI et al., 2009; LOURENÇO; JUNIOR; GARCIA, 2008; FAO, 2013).

Buffalos have economically higher milk production than some breeds of cattle. This is due to the lower cost of production, the greater number of lactating females per year, as well as the animal's hardiness, which makes better use of lower quality fodder, resists the most adverse climatic conditions and diseases (CUNHA NETO, 2003).

This buffalo milk has some peculiarities compared to bovine milk, in particular its sweet taste, due to the high dry extract, evidenced by the higher lactose content (DALMASSO et al., 2011).

Compared to the milk of other species, buffalo milk has higher macro-constituent values. It has a density of between 1.025 and 1.047 g/mL; pH between 6.41 and 6.47; acidity between 0.14 and 0.20 grams of lactic acid per 100 mL of milk (high acidity justified by the high protein content, especially casein); cryoscopy between - 0.531 and - 0.548°C; total solids around 17%, fat between 5.4 and 8%; protein between 3.6 and 5.26%; minerals between 0.79 and 0.83%, of which up to 25% is calcium (BASTIANETTO et al., 2005; COELHO et al., 2004; CUNHA NETO, 2003; FURTADO, 1980).

It is important to note that there is no specific federal legislation to determine the standard of identity and quality of buffalo milk. However, the Secretariat of Agriculture and Supply (SAA) of the state of Sao Paulo has published a resolution valid for the state of Sao Paulo, which established pH values (between 6.40 and 6.90), Dornic acidity (14 to 23 °D), fat (minimum 4.5%), defatted dry extract (minimum 8.57%), density at 15 °C (from 1.028 to 1.034) and cryoscopic index (-0.520 to -0.570 °C) to characterize normal buffalo milk (SÂO PAULO, 1994). It is also important that the sensory and physico-chemical characteristics are evaluated to guarantee products with nutritional value, quality and high acceptability by the consumer market (CUNHA NETO et al., 2005; FARIA et al., 2006).

YOGHURT

Dairy Market

The industrialization of milk and dairy products has become a necessity for the majority of producers *in* Brazil due to the lack of options for marketing it *in natura* and the possibility of higher gross monthly income by adding greater value to the milk produced (GUERRA;NEVES;PENA, 2005; RODRIGUES et al., 2008).

Buffalo milk is widely used to make cheeses and each type requires a specific production technique. When processed from buffalo milk, cheeses generally have a higher yield (NERES et al., 2012; VIEIRA et al., 2011).

In addition to cheese, other dairy products are also produced from buffalo milk, such as yogurt, dulce de leche and even ice cream. Among these products, yogurt occupies a prominent position because, in addition to alleviating the problem of the perishability of milk (whose liquid state and nutritional composition make it prone to microbial degradation), it has a high nutritional value and is highly acceptable to both children and adults. Yogurt is an excellent source of B-complex vitamins, vitamins and minerals, helping to maintain the body, develop strong and healthy bones and teeth, transform carbohydrates, fats and proteins into energy and form and repair body tissues (CUNHA NETO et al., 2005; CUNHA et al., 2008; NOBRE et al., 2006; MATHIAS et al., 2013).

Brazil is one of the largest milk producers in the world, with more than 13 million tons per year, but its per capita milk consumption - less than 100 liters per inhabitant per year - is well below the FAO recommendations (WILKINSON, 2008). Yogurt is one of the few foods that has been known and consumed for more than 4,500 years. Since then, this product has been gaining ground in everyday life, becoming part of many people's eating habits (MORAES, 2004).

Despite the wide variety of flavors and brands available on the Brazilian market, per capita consumption of yogurt in Brazil is only 3 kg per year, which is still small compared to countries such as France, Uruguay and Argentina, where per capita consumption of the product is between 7 and 19 kg per year (BOLINI; MORAES, 2004). However, according to Moraes and Bollini (2010), the per capita consumption of yogurt in Brazil has increased by more than 100% in 20 years, and this increase in consumption is associated with changes in eating habits in Brazilian households.

The total yogurt market in Brazil was worth 2 billion reais in 2005. The light category, which represents an average of 10% of this amount, grew by 11% in 2006 (GAZETA, 2006).

Definition of yogurt

According to Normative Instruction no. 46 of October 23, 2007 of the Ministry of Agriculture, Livestock and Supply (MAPA), yogurt, yoghurt or yoghurt means a product with or without other food substances, obtained by coagulation and lowering the pH of milk, or reconstituted, with or without other dairy products, by lactic fermentation through the action of specific microorganism cultures.

Fermentation is carried out with protosymbiotic cultures of *Streptococcus salivariussubsp. Thermophilus* and *Lactobacillus delbrueckiisubsp. bulgaricus*, which may be complemented by other lactic acid bacteria which, due to their activity, contribute to determining the characteristics of the final product (BRASIL, 2006).

Yogurt has several health advantages for consumers. By reducing the lactose content of milk, as it has undergone a fermentation process, it allows people who are intolerant to this sugar to eat it and benefit from its nutritional properties, such as a higher intake of calcium, more easily digestible protein (casein), as it has already been modified, as well as some very important immunological benefits (KROLOW, 2008).

Buffalo milk yogurt

Yogurt probably originated in the Middle East or India. By always storing milk in the same containers, the nomads selected a microbiota that fermented the milk and produced a food with a pleasant taste (ORDÓNEZ, 2005). Current legislation refers to yogurt as fermented milk which, thanks to the action of *Lactobacillus delbrueckii bulgaricus* and *Streptococcus thermophilus*, has a pasty consistency, a sour taste and smell, a lactic acid content of 0.6% to 1. 5%, a viable lactic microbiota, no impurities, no micro-organisms

pathogens and coliforms (BRASIL, 2007).

The use of buffalo milk for the industrial production of yogurt has shown that the product obtained has different sensory, nutritional and physico-chemical characteristics to the product produced exclusively from bovine milk (AHMAD et al., 2008; BARBOSA et al., 2002; BORGES; MEIDEIRO; CORREIA, 2009; CUNHA NETO et al., 2005; ROCHA et al., 2004; RODAS et al., 2001; VERRUMA-BERNARDI et al., 2006).

Greek yogurt

Among the many special types of yogurt, few authors mention the process for making Greek yogurt, but Varnam and Sutherland (1995) defined the traditional Greek yogurt process (concentrated yogurt), as the product obtained from traditional yogurt, but differentiated by the desorption process in cloth bags, this for small scale and at industrial level by centrifugation. This desorption process makes the yogurt thick and creamy, with a concentration of total solids of approximately 24% and fats of 10%. The original yogurt was obtained from sheep's milk, but it can be obtained from different types of milk.

FRUIT GROWING

General aspects

Brazil is the world's third largest fruit producer, with a production of 43.6 million tons in 2013. Fruit growing is therefore one of the most important sectors of Brazilian agribusiness. Through a wide variety of crops, produced all over the country and in different climates, fruit growing achieves significant results and generates opportunities for small Brazilian businesses (BOLETIM DE INTELIGÊNCIA, 2015).

World fruit production is estimated at 540 million tons, corresponding to US$ 162 billion in commercial value. The harvest of the main producing countries reached 360 million tons in 2004, with China, the leader in fruit production, responsible for 157.7 million tons (ANDIRIGUETO et al., 2007).

In this context, temperate climate fruit production represents only 7.5% (3.02 million tons) and 8% (151,732 ha) of the total area cultivated with fruit trees in Brazil (IBGE, 2009). The fruit-growing sector is one of the main generators of income, employment and rural development in national agribusiness. Fruit-growing has a high income multiplier effect and is therefore strong enough to boost stagnant local economies with few development alternatives (BUAINAIN;BATALHA, 2007). In 2008, Brazil exported approximately 888,000 tons of fruit, worth U$S 723 million, of which temperate climate fruit accounted for 197,964 tons and U$S 267,382 million (FAOSTAT, 2011).

In 2011, Brazil harvested 45.1 million tons, 7.1% more than the previous year, when the volume harvested was 42.1 million tons. According to data from the Foreign Trade Secretariat of the Ministry of Development, Industry and Foreign Trade, 682,000 tons of fresh fruit were exported in 2011, earning US$ 634.5 million, with melons, grapes, mangoes, apples, lemons and bananas being the main fruits exported (SEAB, 2012).

Fruit storage

The storage capacity of fruit and vegetables depends on various metabolic reactions modulated by temperature, transpiration and the concentration of gases in the atmosphere, such as CO_2, O_2 and ethylene. Respiration is the main physiological process involved in the post-harvest physiology of vegetables and fruit (CALBO et al., 2007). The post-harvest quality of fruit is related to minimizing the rate of deterioration, i.e. maintaining firmness, color and appearance, in order to keep them attractive to the consumer for a longer period of time (CHITARRA; CHITARRA, 2005).

One of the ways in which cold is used in food preservation is refrigeration. In this process, the temperature of the food is reduced to between -1 and 8° C, which implies changes in the sensitive heat of the product. In this way, it is possible to reduce the speed of microbiological and biochemical transformations in food, thus prolonging its useful life for days or weeks (TOLEDO, 1991; FELLOWS, 2006). This type of processing does not improve the quality of the products, so only healthy, quality tissues should be refrigerated, since the low temperature does not destroy the pathogen, but only reduces its activity (ORDÓNEZ, 2005).

Another way of preserving food at low temperatures is by freezing. In this method, part of the water in the food changes its state, forming ice crystals (FELLOWS, 2006). In this way, the water activity of the food is reduced, which increases the shelf life of the product. Freezing slows down, but does not stop, the physical and biochemical reactions that lead to food spoilage, and during frozen storage there is a slow and progressive change in the sensory quality of food products (RAHMAN; RUIZ, 2007).

Optimum storage temperatures vary from product to product, and it is very important to select the temperature for each product handled. Tropical and subtropical crops are sensitive to cold storage and show damage when subjected to these temperatures. Cold damage occurs in various ways, such as surface depressions, internal discoloration, tissue collapse, increased susceptibility to disease and reduced quality (PINTO; MORAES, 2000).

Sour cherry tree (Eugenia uniflora L.).

The species *Eugenia uniflora* L. is native to the region that stretches from Central Brazil to Northern Argentina. However, as it is a species that adapts easily, it has spread throughout almost the whole of Brazil, as well as in various parts of the world (SANCHOTENE, 1989; BEZERRA; SILVA JUNIOR; LEDERMAN, 2000). In terms of its wide geographical distribution and diversity of ecosystems, the pitangueira occurs predominantly in the Atlantic rainforest as a tree that reaches between 4m and 5m in height, and can rarely reach 8m to 12m, or even as a 0.70m shrub in Restinga areas (SALGUEIRO et al., 2004).

The fruits are globose berries, crowned by a persistent calyx, with flattened poles and seven to eight longitudinal grooves. When the ripening process begins, the epicarp turns from green to red and from there to almost black (SANCHOTENE, 1989). However, some plants have orange or red fruits, even when they have reached maturity, usually with one to two seeds, sporadically three to four and rarely more than this (ANDERSEN; ANDERSEN, 1988). On average, ripe fruits have a yield (edible portion) of 65 % and 80 % (SANTOS et al., 2002) for the red and purple types, respectively, and Bezerra, Silva Junior and Lederman, (2000) found values ranging from 74.6 to 88.4 %.

In the myrtle family specifically, to which the pitanga belongs, the fruits generally fall into the class of fleshy, juicy fruits (GEMTCHÜJNICOV, 1976). According to Franco (2006), pitanga has 6.40% carbohydrates, 1.02% protein and 1.90% lipids. The Brazilian Table of Food Composition (TACO, 2006) shows 10.2% carbohydrates, 0.9% protein, 0.2% lipids, 88.3% moisture and 0.4% ash.

Figure 1 - Sour cherry tree (*E. uniflora* L.): (A) Tree; (B) Leaves; (C) Green fruit; (D) Fruit at different stages of ripeness.

Source: Authors, 2018.

Sour cherries are consumed fresh or in the form of soft drinks and juices, ice cream, sweets, liqueurs, wine and jellies (OLIVEIRA, 2006; LIMA et al., 2002), thus increasing the interest of producers and consumers (KUSKOSKI et al., 2006).

BIOACTIVE COMPOUNDS: ASCORBIC ACID AND ANTOCIANINAS

Ascorbic acid

Vitamins are organic compounds required in minimum quantities to promote growth, maintain life and the ability to reproduce. The daily intake of vitamins needed to ensure the proper functioning of the body is specified as the Recommended Daily Allowance (RDA) (RIBEIRO; SERAVALLI, 2007). Vitamin C is the common name given to 2,3-enediol-L-gulonic acid, which is a powerful antioxidant because it prevents oxidation, i.e. the loss of electrons. The molecules of ascorbic acid (vitamin C) undergo oxidation before other molecules oxidize, preventing and protecting these other molecules from oxidation (PEREIRA, 2008).

Ascorbic acid occurs only in plant tissue (where its function is unknown) and is also synthesized by almost all mammals, so it is not an essential vitamin for them, except for primates, guinea pigs and some vegetarian bats. The latter use fruit as a source of the vitamin (COULTATE, 2004).

Ascorbic acid is a water-soluble compound that corresponds to an oxidized form of glucose, C6H8O6 (176.13 g/mol), and is a six-carbon alpha-ketolactone, forming a five-membered lactone ring and a bifunctional enadiol group with an adjacent carbonyl group (Figure 2). It cannot be synthesized by humans or primates (VANNUCCHI; ROCHA 2012).

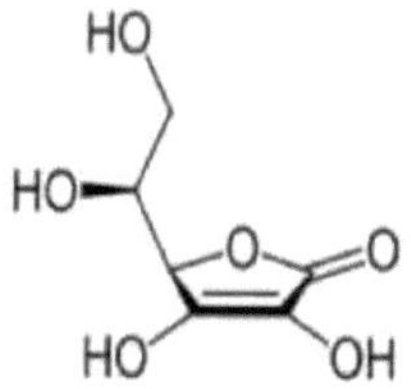

Figure 2 - Structural formula of vitamin C. Source: Teixeira Neto, 2009.

This vitamin belongs to an organic group called Iactones, which are carboxylic acids that are transformed into cyclic esters, i.e. closed-chain esters that have spontaneously lost water. Its polar molecule with four hydroxyls (OH), two of which are in the C=C position, can interact with each other through hydrogen bridges, resulting in an increase in the acidity of vitamin C, which has good water solubility (PEREIRA, 2008).

Vitamin C is easily degraded, mainly by aerobic or anaerobic oxidation, both of which lead to the formation of furaldehydes, compounds that polymerize easily, forming dark pigments. It is also destroyed by light and heat (BOBBIO; BOBBIO, 1992). It is therefore very susceptible to oxidation in aqueous solution by enzymatic and non-enzymatic processes, especially when catalyzed by transition metal ions such as Cu2+ and Fe3+. Heat and light also accelerate this process, while factors such as pH, oxygen concentration and water activity greatly influence the reaction speed. As the hydrolysis of dehydroascorbic acid occurs very easily, the oxidation of hydroascorbic acid often represents a fundamental step in limiting the oxidative degradation of vitamin C (DAMODARAN; PARKIN; FENNEMA, 2010).

Anthocyanins

Anthocyanins belong to the group of flavonoids, a group of natural pigments with varied phenolic structures (NIJVELDT et al., 2001; KUSKOSKI et al., 2004). They are the components of many red fruits and dark vegetables and are highly concentrated in the skins of dark grapes (DOWNHAM et al., 2000). They play a significant role in preventing or delaying the onset of various diseases due to their antioxidant properties (MARTÌNEZ-FLÓREZ et al., 2002; KUSKOSKI et al., 2004; DOWNHAM et al., 2000).

Anthocyanins are glycosides whose chemical structure includes a sugar residue in position 3, which is easily hydrolyzed by heating with 2N HCl. The product of this hydrolysis is the glycoside component and the aglycone, called anthocyanidin (ARSEGO et al., 2002).

The most common anthocyanidins are: pelargonidin, cyanidin, delphinidin, peonidin, petunidin and malvidin (Figure 4). They have in common a hydroxylation of the flavylium cation in the C-3, C-5 and C-7 positions, while they differ in the substitution pattern of the hydroxyl and/or methoxyl groups in the B ring (MAZZA, 2007; KOH; MITCHELL, 2006).

The most widely described property of anthocyanins is their antioxidant action. Cells and tissues in the human body are continually suffering damage caused by free radicals and reactive oxygen species, which are produced during normal oxygen metabolism or are induced by exogenous damage (MARTÌNEZ-FLÓREZ et al., 2002; NIJVELDT et al., 2001; WANG et al., 1999).

In addition, anthocyanins are a viable alternative for providing red color to foods from natural sources (SARNI-MANCHADO, 1996). They are soluble in water, which facilitates their incorporation into aqueous systems. According to Ozela, (2004) anthocyanins can replace the artificial colorants red 40, ponceau 4R, erythrosine and bordeaux S.

The use of anthocyanins as a coloring agent is indicated for foods that are not subjected to high temperatures during processing, with a short storage time and packaged in such a way that exposure to light, oxygen and humidity is minimized (FENNEMA, 2000), which is why new studies are needed in the search for viable sources that have better stability and low cost, as well as an increase in their coloring power (PAZMINO-DURAN, 2001).

CHAPTER II

MATERIAL AND METHODS

The research was carried out at the Food Technology Laboratory of the State University of Parà, Salvaterra, Marajó, Parà.

Obtaining raw materials

The fruits were collected in the municipality of Salvaterra, Marajó Archipelago, Parà, in the early hours of the morning. They were hand-picked from the trees (Figure 3), totaling 100 fruits which were packed in plastic boxes to avoid mechanical damage, as they are a sensitive fruit. They were transported to the Food Technology laboratory at the State University of Parà, Salvaterra Campus, where they underwent a selection process, separating those that were damaged from those that were healthy. After this process, the fruit was sanitized in 10 ppm chlorinated water for 10 minutes.

Figure 3 - Sour cherry trees (*Eugenia uniflora* L.).

Source: Authors, 2018.

Raw materials for yogurt production

The raw materials used to make Greek buffalo milk yogurt with sour cherry pulp syrup were:

- ✓ Fresh buffalo milk;
- ✓ Sour cherry pulp;

- ✓ UHT whole milk - CCGL®;
- ✓ Refined sugar - UNIÂO®;
- ✓ Freeze-dried milk culture - Bòrico®.

Obtaining raw materials

The buffalo milk used in the yogurt production process was purchased from the Mironga farm, located in the municipality of Soure, in the state of Parà. The milk purchased was of good physical and chemical quality for the production process.

The sour cherry pulp (*Eugènia Uniflora* L.) was purchased in the municipality of Salvaterra in the state of Parà and harvested by hand during the harvest season.

The whole milk and sugar were purchased in the municipality of Salvaterra, in the state of Parà, from local businesses and both condiments were within their expiration dates and stored properly.

The freeze-dried milk culture used for the fermentation process, made up of *Lactobacillus acidophilus* and *Streptococcus subsp.Thermophilus* bacteria, was purchased in the municipality of Belém, Parà, from local businesses.

Making syrup from the pulp of sour cherry fruit (Eugenia Uniflora L.)

To make the sour cherry pulp syrup in the desired proportions of 10, 20 and 30% (obtained in preliminary tests), sugar, fruit pulp and water were used to complete 600 grams of natural yogurt of each formulation according to the concentrations in Table 1.

Table 1 - Formulations for making sour cherry syrups to enrich Greek yogurt.

Raw materials	**Formulations**		
	10% syrup	**20% syrup**	**Syrup 30%**
Pulp (%)	50,06	51,77	53,60
Sugar (%)	33,01	35,29	37,43
Water (%)	13,39	12,94	12,51

To produce the syrups, the pulp, sugar and water were homogenized and cooked at 80 °C/15 min, enough time for the syrup to acquire the consistency needed to be homogenized with the yoghurt masses according to preliminary tests.

Greek yogurt made from buffalo milk with sour cherry syrup (Eugenia Uniflora L.)

The following concentrations were used for the production of buffalo milk Greek yogurt with sour cherry syrup (Table 2).

Table 2 - Concentrations of raw materials for the production of Greek buffalo milk yogurt with sour cherry syrup.

Raw materials	Percentage value
Bovine milk (%)	94,33
Sugar (%)	3,30
Whole milk powder (%)	2,37

The processing of Greek yogurt from buffalo milk (as a dairy base) with the addition of different concentrations of sour cherry pulp is shown in Figure 4.

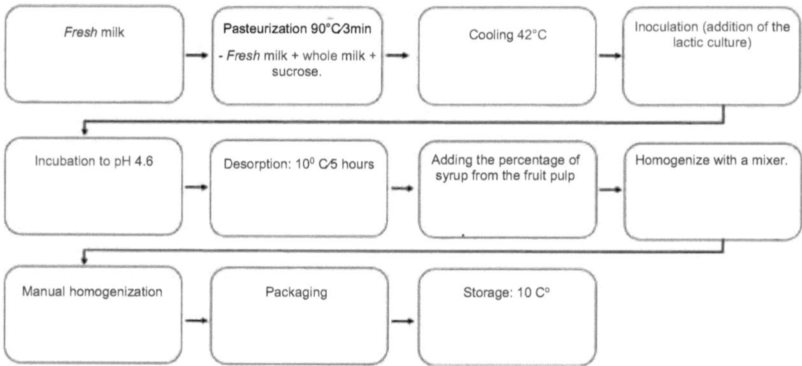

Figure 4 - Flowchart of the production process for Greek buffalo milk yogurt with sour cherry syrup.

The fresh buffalo milk was pasteurized in a previously sanitized container to avoid possible contamination. The milk was pasteurized at 90° C for a period of 3 minutes, with the powdered milk and sucrose previously weighed according to Figure 5.

Figure 5 - Pasteurizing buffalo milk at 90° C for 3 minutes.

After pasteurization, the milk was cooled to a temperature of 42° C for the inoculation of the lactic culture, 400 grams per 1 liter of milk, according to the manufacturer (Bio Rich®). After inoculating the milk, 10 mL aliquots were separated into beakers so that the pH could be monitored until it reached (4.6) in a BOD oven (BIOFOCO® brand) and then measurements were taken every 1 hour using a handheld pH meter to monitor the fermentation, which lasted an average of 6 hours, as can be seen in Figures 6 and 7.

Figure 6 - Start of the fermentation process of buffalo milk.

Figure 7 - After the fermentation process.

At the end of the fermentation process, the container with the yogurt went through a cooling process in a cold bath in order to inactivate the fermentation process of the lactic acid-producing microorganisms. The yoghurt mass was then transferred to sterilized cotton bags for draining at 10 °C for 5 hours, as can be seen in Figure 8.

Figure 8 - Desiccation process (A), Yogurt after desiccation (B).

After desorption, the yogurt mass was weighed and divided according to the formulations defined and then added to the percentage of sour cherry pulp syrup and homogenized in a mixer (ARNO®) and then manually as shown in Figure 9.

Figure 9 - Greek yogurt made from buffalo milk with sour cherry syrup.

PHYSICAL AND CHEMICAL ANALYSIS

For the physico-chemical characterization of the sour cherry fruit and the Greek yoghurts made from buffalo milk and sour cherries, the following analyses were carried out: pH, titratable acidity, humidity, ash, soluble solids (°Brix), reducing sugars, lipids, density, ascorbic acid, total anthocyanins.

Ascorbic acid

The ascorbic acid determination analysis was carried out using the iodine oxidation method (Iodimetry), where a standard ascorbic acid solution was prepared by adding 5 drops of starch solution as an indicator and then titrating 50 mL of the solution with 2% iodine. The same process was carried out for the samples by weighing 5 g and diluting them in 50 mL of distilled water according to Silva et al. (1995).

Total anthocyanins

The sample was weighed between 0.400 and 0.420g, diluted 10 times and the volumes of pH 1.0 and pH 4.5 buffer solutions were added. The solution was then filtered through filter paper to remove any interfering solids and read in duplicate on a spectrophotometer, according to Askar and Treptow (1993) adapted by Rogez (2000) expressing the result as mg of cyanidin-3-glucoside per 100 g of fresh mass.

MICROBIOLOGICAL ANALYSIS

The microbiological analyses were carried out at the Food Laboratory of the University of Parà, Salvaterra campus. In order to verify the microbiological quality of the raw material and the yogurt formulations prepared, analyses were carried out in accordance with Normative Instruction No. 46 of October 2007 to determine Coliforms at 35 °C, Coliforms at 45 °C by plating on 3M™ Petrifilm plates and Staphylococcus Aureaus A.O.A.C (2003). Where the 25 g samples of each were diluted in 1% peptone water, and then from these dilutions the other 10^{-1} and 10^{-2} were made and then 1 mL of each dilution was incubated in their respective analysis plates. After 46 hours, the plates were incubated at 35° C and the results were interpreted.

STATISTICAL ANALYSIS

The results of the analyses were evaluated using ANOVA analysis of variance and the TUKEY test at 5% probability in the Assistat 7.1 statistical program and the Statistica 7 statistical program.

CHAPTER III

RESULTS AND DISCUSSION

Physico-chemical characterization

The results of the physico-chemical analysis of the sour cherry fruit at the green and ripe stages can be seen in Table 4.

Table 4 - Physical and chemical characterization of sour cherry fruit at the green and ripe stages.

Parameters	**Fruits**		
	Green	**Maduro**	**BRAZIL (2000)**
PH	$3.31^{b} \pm 0.00$	$3.53^{a} \pm 0.00$	2,5 a 3,4
humidity (%)	$87.26^{a} \pm 0.16$	$85.88^{a} \pm 0.62$	-
Titratable acidity (%)	$0.34^{a} \pm 0.02$	$0.04^{b} \pm 0.00$	Min 0.92
Soluble Solids (°Brix)	$11.0^{a} \pm 0.00$	$11.0^{a} \pm 0.00$	Min 6.0
SS/ATT ratio	32,35	275	-
Ash (%)	$0.34^{a} \pm 0.03$	$0.43^{a} \pm 0.00$	-
Anthocyanins (mEqgCyanidin-3-glycoside/100g)	*nd	$179.270^{a} \pm 1.80$	-
Ascorbic Acid (mg/100g)	$109.14^{b} \pm 0.09$	$124.076^{a} \pm 0.23$	-

The pH values found for sour cherries were 3.31 and 3.53 for unripe and ripe fruit, respectively. These values were close to those found by Bagetti (2009), who in his study found a pH value of 3.38. Mélo, Lima and Nascimento (2000) found pH values of 2.08 for ripe fruit and 2.09 for green fruit, which were lower than those found in this study. Studies carried out on gabiroba fruit (Campomanesia xanthocarpa B.), a species from the same family, found higher values of 3.89 (SANTOS et al., 2009).

Table 3 shows that there was a significant difference, at $p<0.05$, in the pH of pitanga between the green and ripe samples. It is worth noting that of the parameters studied, only the pH was outside the standards in force for sour cherry fruit pulp (BRASIL, 2000).

For humidity, the values found did not differ significantly, being 85.88 and 87.26 %, respectively, in the ripe and green stages. These values are higher than those of Bagetti, (2009) who found a value of 76.8 % for pitanga fruit, but in the studies by Barreto, (2011) his value is higher, being 90.42 % in ripe pitanga fruit. Approximate values were found for acerola 92.4%, Cambuci 88.8% and cherry 90.71% (FREITAS et al., 2006; VALLILO et al., 2006; CAMLOFSKI, 2008).

It can be seen that the titratable acidity values differ, at $p>0.05$, with the average values obtained being 0.044 and 0.347%, respectively, for ripe and unripe fruit and during the ripening of the fruit it is possible to see that the acidity value decreases. Barreto, (2011) found 1.62%, close to Bagetti, (2009) who found 67% in his studies on sour cherries, but lower than the present study on ripe fruit and that the values found are below those allowed by legislation.

The soluble solids content (°Brix) did not differ significantly, with both ripe and unripe fruit remaining in the range of 11.00 °Brix, values very close to those of Bagetti, (2009) which was 11.5 °Brix and higher than the result of the study by Barreto, (2011) which obtained a value of 8.29 °Brix. In the study by Lopes, Mattieto and Menezes, (2005) they found higher values in pitanga fruit from Vanilhos, Sao Paulo 12.29 °Brix and the soluble solids content was within the limits allowed by the legislation BRASIL, (2000).

According to the results for the SS/ATT ratio, the values differed significantly when compared using the Tuckey test. In the consumer market for fresh and/or processed fruit, a high TSS/TTA ratio is desirable (AGUIAR, 2006). It is worth noting that the relationship between the acidity of the fruit and its sweetness is directly linked to the stage of ripeness of the fruit, because if the ratio is low it indicates high acidity of the fruit, as is the case with the ratio of 32.35 for green stage fruit and 275 for ripe fruit, the higher this value the greater the sweetness. Teixeira (2000) reported an increase in this ratio as the fruit matured, obtaining 43.87 in green-skinned bacuri and 56.84 in yellow-skinned bacuri. The values for ash also did not differ significantly, standing at 0.30 and 0.34 % for ripe and green fruit respectively, close to Taco's (2006) value of 0.4 %. Oliveira, Figuêiredo and Queiroz, (2006) found levels of 0.22 in pitangas from the region of Campina Grande, Paraiba State.

The ascorbic acid content of the ripe and unripe fruit was 124.076 mg/100g and 109.14mg/100g, which were significantly different from each other. Santos et al. (2009) found 233.56±11 mg/100g of vitamin C in gabiroba fruit, Santos et al. (2002) found lower levels than this study in pitangas of both varieties (33.00 and 38.35 mg/100g, red and purple, respectively). Lower values were also found by Oliveira, Figêiredo and Queiroz, (2006) of 13.42 mg/100g in fruit from the region of Campina Grande, Paraiba.

The anthocyanin content of ripe pitanga fruit was 124.076 mg of cyanidin-3-glucoside/100g and no anthocyanins were detected in unripe fruit. The appearance of this compound is related to the degree of ripeness of the fruit. Lima et al. (2000) found anthocyanin values of 22.50 mg/100g for this ripe fruit, lower than in this study. Lima, Mélo and Lima, (2005) also found lower values of 16.23 mg.100g-1 for purple pitanga fruit.

According to Martinez-Flórez et al. (2002), Nijveldt et al. (2001), and Wang et al. (1999) cells and tissues of the human organism are continuously suffering aggressions caused by free radicals and reactive oxygen species, which are produced during normal oxygen metabolism or are induced by exogenous damage. Anthocyanins therefore play a significant role in preventing or delaying the onset of various diseases due to their antioxidant properties (MARTÌNEZ-FLÓREZ, 2002; KUSKOSKI et al., (2004); DOWNHAM et al., (2000).

Physico-chemical and microbiological characterization of milk

Table 5 shows the results obtained for the physico-chemical characterization of buffalo milk for the parameters pH, moisture, acidity, lipids, density, ash and total solids.

Table 5 - Physical and chemical characterization of buffalo milk produced in Soure.

Parameters	Milk	*Parameter (SAA)	**Parameter (bovine milk)
Ph	6,85±0,080	6,40 a 6,80	-

Humidity (%)	83,2±0,568	-	86,50
Acidity (% lactic acid)	0,129±0,000	0,14% a 0,23	0,78 a 1,21
Lipids (%)	9,00±0,05	Minimum 4.5	Minimum 3.0
Density (g/L)	1.056,58±0,08	1,028 a 1,034	1,028 a 1,034
Ash (%)	0,5±0,00	Maximum 8.4	0,73
Total Solids (%)	16,8±0,03	-	-

*Parameters of the São Paulo Agriculture and Supply System; ** BRAZIL (2011).

Due to the lack of specific legislation for buffalo milk, the results were compared with those described by the SAA (Secretaria de Agricultura e Abastecimento de Sâo Paulo) and with the parameters for bovine milk. The pH value (6.85) was found to be in line with that permitted by the SAA, which allows a range of 6.40 to 6.80. These results are in line with the studies by Borges, Medeiros and Correia, (2009) and Guerra, Neves and Pena, (2005), respectively 6.65 and 6.8.

The moisture content was 83.2 %, which is slightly below the norm for bovine milk, where the moisture content is 86.50 %. However, these variations are largely due to entropic issues and the difference between the breeds themselves, taking into account that buffalo milk has a higher solids content than bovine milk. According to Oliveira et al. (2008), their study of buffalo milk found a total solids content of 20.1%.

The acidity value of the buffalo milk, which was 0.12 %, was below that established by the SAA (0.14 % to 0.23 %) and bovine legislation (0.78 % to 1.21 %) in lactic acid. The fat content of 8.4% was higher than the minimum values set by the SAA and BRASIL, (2011), thus showing that the milk is within the legal standards stipulated by the São Paulo state supply department. This value is higher than that found by Cunha Neto et al. (2005) where they found an average of 6.82 %, Oliveira et al. (2009) found a value of 8.6 % and Coelho et al. (2004) found a value of 6.8 % for buffalo milk.

The density of the buffalo milk of 1,056.58g/m3 was not in accordance with the SAA nor with the legislation for bovine milk BRASIL (2011), Cunha Neto et al., (2005) found an average value for density of 1,033.84 g/L in their studies with buffalo milk. The ash content of 0.5 % is lower than that established by the SAA and lower than the limits for bovine milk. Guerra, Neves and Pena (2005) found a value of 4.0 % for ash. For total solids, the value found was (16.8 %), the same as in the study by Lopes et al. (2009) who found a value of (16.8 %), Oliveira et al. (2009) a value of (20.1 %) and Coelho et al. (2004) (17.2 %), a value close to that of Furtado (1980) (17.09 %) and Macedo et al. (2001) (17.01 %), but Nader Filho et al. (1984) mentioned a content of 19 % for buffalo milk.

Table 6 shows the results of the microbiological analyses carried out on fresh buffalo milk purchased in Soure.

Table 6 - Microbiological analysis of fresh buffalo milk.

	Analysis		
	Coliforms at 35°C (UFG/g)	**Coliforms at 45°C (CFU/g)**	**Staphylococcus Aureus (CFU/g)**
***Fresh* milk**	$3{,}0x10^3$	<3	1,5x104
***Standards**	Max. 1x104	Max. 1x104	Max. 1x104

*BRAZIL, 2002.

According to the results obtained from the microbiological analysis of the milk in natura, current legislation BRASIL (2002) allows a count in CFU/g of up to $1x10^4$ colonies, and it can be seen that only *Staphylococcus Aureus* showed counts above the permitted level, indicating some fault during the milk handling process. Coliforms are indicators of contamination of the environment and feces (BRITO;BRITO;PORTUGAL, 2002), showing the important contribution of milking hygiene to the total bacterial count of milk.

Physico-chemical characterization of sour cherry Greek yogurt

Table 7 shows the results for the physical-chemical characterization of Greek yogurt formulations with different concentrations of sour cherry syrup in terms of pH, moisture, acidity, soluble solids, insoluble solids, lipids, reducing sugars and ash.

Table 7 - Physico-chemical characterization of Greek yogurt with different concentrations of sour cherry syrup

Parameters	**Formulations**

	***SA	10%	20%	30%
pH	4.13^{a} ±0.05	3.90^{b} ±0.00	3.63^{c} ±0.05	3.63^{c} ±0.05
***Acidity (%)**	0.39^{c} ±0.01	0.39^{c} ±0.01	0.63^{a} ±0.00	0.59^{b} ±0.00
Humidity (%)	68.17^{a} ±0.04	60.30^{b} ±0.01	58.07^{c} ±0.00	58.26^{c} ±0.01
****°Brix**	15.00^{b} ±0.00	24.00^{a} ±0.00	24.00^{a} ±0.00	24.00^{a} ±0.00
Lipids (%)	42.49^{a} ±0.01	39.85^{b} ±0.11	19.88^{d} ±0.63	23.37^{c} ±0.21
******AR (%)**	4.7^{b} ±0.90	4.8^{b} ±0.60	4.9^{ab} ±0.12	6.9^{a} ±1.01
Ash (%)	0.5^{a} ±0.02	0.7^{a} ±0.03	1.2^{a} ±0.22	0.5^{a} ±0.03

Different letters in the same column indicate significant differences according to the Tukey test (p< 0.05). Acidity (% lactic acid); ** °Brix - Soluble solids (%); ***SA - No syrup added; **** Reducing sugars.

As for the pH parameter, it can be seen that the SA formulation differs statistically from the others according to the Tuckey test (p<0.05), with an average of 4.13. The same is true for the 10% formulation (3.90), which differs statistically from the formulations with 20% and 30% sour cherry syrup, which do not differ from each other with respective values of 3.63 and 3.63. Neves (2007) found a pH value of 4.6 for açai yogurt, which is higher than that found in this study.

The acidity values also differed for the formulations with 20% and 30% sour cherry syrup, which had higher values of 0.63% and 0.59% respectively. There was no significant difference between the SA and 10% sour cherry syrup formulations. However, the percentage of acidity of the SA and 10% samples was outside the range recommended by the legislation, with only the 20% and 30% sour cherry syrup formulations being within the range, with values very close to the legislation for fermented milks, which provides for values between 0.60% and 2.00% lactic acid BRASIL (2007).

The titratable acidity value found is lower than that reported by Cunha Neto et al. (2005) and Borges et al. (2009), which averaged 1.13 and 0.98 g/100g, respectively, for yogurt made from buffalo milk.

The moisture content showed a significant difference between the SA and 10% formulations, with values of 68.17% and 60.30%, respectively. Medeiros et al. (2007)

analyzed commercial brand yogurts and found a value of 80.53%, higher than the sour cherry Greek yogurt produced in this study. With regard to soluble solids, only the SA formulation showed a significant difference from the others, with a value of 15.0. However, the others did not differ from each other, with the respective values for the 10%, 20% and 30% formulations being (24.00). Oliveira et al. (2008) obtained 15.30 to 18.20°Brix for araticum yogurt with different concentrations of the fruit, these values being lower than the formulations with the fruit syrup added in this study.

The lipid levels found for the formulations differed statistically, being 42.49, 38.85, 19.88 and 23.37%, respectively, for SA, 10%, 20% and 30% sour cherry fruit syrup, however, both formulations were in accordance with the legislation for Greek yogurt, which recommends standards of at least 3% (BRASIL, 2007). The value of reducing sugars for the SA, 10% and 20% formulations did not differ statistically with values of 4.7, 4.8 and 4.9%, respectively, but the 20% and 30% formulations did not differ from each other with respective values of 4.9% and 6.9%, but both differed from the others according to the Tuckey test ($p<0.05$). In their studies, Borges, Medeiros and Correia (2009) found an AR value of 9.47% for yogurt made from buffalo milk and 7.27% for bovine yogurt. These values are higher than the sour cherry flavored Greek yogurt produced in this study.

Although there is no legislation on ash content, the values found did not differ significantly from each other, with an average of 0.7%. Medeiros et al. (2007) analyzed commercial brand yogurts and found 0.46% ash, which was lower than the average found in this study. This parameter is not covered by current Brazilian legislation (BRASIL, 2000) and is deeply influenced by the raw material, which may be a consequence of the high ash content of buffalo milk (VERRUMA; SALGADO 1994).

Table 8 shows the results of the analysis of ascorbic acid (AA) and anthocyanins in Greek yoghurts made from buffalo milk with different concentrations of sour cherry syrup.

Table 8 - Ascorbic acid and anthocyanin content of buffalo milk Greek yogurts with different concentrations of sour cherry syrup.

Parameters	Formulations *AS	10%	20%	30%
Asc. Ascorbic (mg/100g)	25.66^{c} ±0.77	113.38^{b} ±0.85	122.59^{b} ±8.44	141.93^{a} ±0.75
Anthocyanins (mEqgCyanidin-3-glycoside/100g))	**nd	39.399^{C} ±0.28	49.806^{b} ±0.00	53.770^{a} ±0.05

Different letters in the same column indicate significant differences according to the Tukey test ($p < 0.05$).*SA - No syrup added; **nd - not detected.

According to the results for the ascorbic acid (AA) content, only the SA and 30% formulations showed a significant difference with respective values of 25.66 mg/100g and 141.93 mg/100g. With regard to the 10% and 20% formulations, there was no significant difference with values of 113.38 mg/100g and 122.59 mg/100g respectively. Souza (2011), in his study of yogurt made with Amazonian pulps, found a range from 140.22 mg/100g to 240.40 mg/100g. Data on vitamin C in dairy products is scarce, but Favaro-Trindade et al. (2006) reported values of 147.7 mg/100g in fermented acerola ice cream, a value close to that of the 30% formulation and higher than the others. However, it can be seen that all the formulations had a vitamin C content above 60 mg/100mL, the recommended daily intake for healthy adults (BRASIL, 2005).

For the concentration of anthocyanins, the SA formulation showed no concentration of this pigment and the other formulations with 10, 20 and 30% sour cherry syrup showed significant differences with values of (39.39 mg/100g), (49.80 mg/100g) and (53.77 mg/100g) after the statistical test.

The concentrations found for the Greek yogurt formulations are lower than the average value found for the pulp, which was 179.27 mg/100g. This is because the syrups had to be cooked at a temperature of 80 °C±2 for 15 minutes, which may have influenced the anthocyanin content since these are thermolabile compounds that degrade under the action of high temperatures. Ribeiro et al, (2011) found values of (33.3 mg/100g) of anthocyanins in açai pulp and (235.8 mg/100g) in juçai pulp. Costa et al. (2012), when developing yogurt with juçai pulp, concluded that due to the addition of pulp from the fruit of the Euterpe edulis Martius palm tree, which is rich in nutrients and antioxidants, their final product would also be a good source of anthocyanins and minerals.

Microbiological characterization of Greek yogurt

Table 9 shows the results of the microbiological analyses carried out on yogurts with different proportions of sour cherry pulp syrup, analyzed for coliforms at 35 °C, 45 °C and *Staphylococcus aureus*.

Table 9 - Microbiological analyses for Greek buffalo milk yogurt with different concentrations of sour cherry syrup.

Formulations	Analysis		
	Coliforms at 35°C (CFU/g)	**Coliforms at 45°C / (CFU/g)**	***taphylococcus aureus* (CFU/g)**
***AS**	≤3	≤3	≤3
10%	≤3	≤3	≤3
20%	≤3	≤3	≤3
30%	≤3	≤3	≤3
****Legislation**	10 NMP/mL[2]	1.0×10^1 NMP/mL	-

*IN No. 46 of 2007 - indicative samples (BRASIL, 2007).

According to Table 9, the results show the absence of micro-organisms in the prepared formulations, indicating values lower than those recommended by current legislation. Based on these results, it is possible to indicate that the products were within the microbiological standards, thus demonstrating that they were prepared following good manufacturing practices for food products.

CONCLUSION

The buffalo milk and sour cherry fruit had good physical and chemical characteristics, and the fruit was found to contain high levels of ascorbic acid and anthocyanins when compared to studies by other authors on the same fruit. The processing of Greek buffalo yogurt with the addition of sour cherries is ideal for adding ascorbic acid and anthocyanin values to the product, and in the formulation with 30% syrup the ascorbic acid and anthocyanin values were the highest, with ascorbic acid being above the recommended daily intake.

REFERENCES

AGUIAR, L. P. **Quality and potential use of bacuris from the Mid-North region.** 2006. 122 f. Dissertation (Master's Degree in Food Technology) - Federal University of Ceará. 2006.

AHMAD, S.; GAUCHER, I.; ROUSSEAU, F.; BEAUCHER, E.; PIOT, M.; GRONGNET, J. F.; GAUCHERON, F. Effectsofacidificationonphysic-chemicalcharacteristicsofbuffalomilk: A comparisonwithcow'smilk. **Foodchemistry**, v. 106, p. 11-17, 2008.

AJAIKUMAR, K. B. et al. **The inhibitionofgastricmucosalinjuryby Punica granatum L. (pomegranate) methanolicextract.J. Ethnopharmacol.** Lausanne,v. 96, p.171-76, 2005.

ANDERSEN, O.; ANDERSEN, V. U. **As Frutas Silvestres Brasileira**. Rio de Janeiro: Globo Rural, p. 203, 1988.

AOAC Official Method. **Enumeration of Staphylococcus aureus in Selected Meat, Seafood, and Poultry 3MTM PetrifilmTM Staph Express Count Plate Method**.2003.

ANDRIGHETTO, C. The buffalo milk production chain: a university view. In: SYMPOSIUM OF THE BUBALIN PRODUCTION CHAIN, 2., 2011 AND INTERNATIONAL SYMPOSIUM OF BUFFALO PRODUCTION CHAIN, 1., 2011, Botucatu. **Proceedings...** Botucatu: FEPAF, 2011.

ARSEGO, J. L. et al. Kinetics of anthocyanin extraction in raspberry (*Rubus idaeus*) and blackberry (*Rubus fructicosus*) fruits. In: **XVI Congresso Brasileiro de Fruticultura**. 2002.

ASKAR, A.; TREPTOW, H. **Qualityassurance in tropical fruitprocessing**. New York: Springer-Verlag, p. 231, 1993.

BRAZILIAN ASSOCIATION OF BUFFALO BREEDERS - ABCB. **Technical Regulations for the ABCB Purity Seal Program**. 2016. Available at: < http://www.bufalo.com.br/abcb.html >. Accessed on: 01 Nov. 2018.

ASSUMPÇÂO, J.; NASCIMENTO, M. T. Structure and floristic composition of four vegetated restinga formations in the Grussai/Iquipari lagoon complex, Sao Joao da Barra, RJ, Brazil. **Acta botânica brasil,** v. 14, p. 301-315, 2000.

AURiCCHiO, M. T.; & BACCHi, E. M. Leaves *of Eugenia uniflora* L. (pitanga: pharmacobotanical, chemical and pharmacological properties. **Revista Instituto Adolfo Lutz.** v. 62, n. 1, p. 55-61, 2003.

BAGETTI, M. **Physico-chemical characterization and antioxidant capacity of pitanga (*Eugenia uniflora L.*)**. 2009. 84 f. Dissertation (Master's Degree in the Food Science and Technology Postgraduate Program)- Federal University of Santa Maria, Rio Grande do Sul, 2009.

BARBOSA, R. A. et al. Formulation and elaboration of yogurt from frozen and partially skimmed buffalo milk (Bubalusbubalis). **Revista do Instituto de Laticinios "Càndido Tostes"**, v. 57, n. 324, p. 31-34, 2002.

BARRETO, I. M. A. **Characterization of purple pitanga (*Eugenia uniflora*) pulp dehydrated in a foam bed**. Itapetinga, Bahia, 2011. Originally presented as a dissertation for the postgraduate program in food engineering, State University of Southwest Bahia, 2011.

BERNARDES, O. Buffaloes in Brazil. In: II SIMPÒSIO DE BÙFALO DE LAS AMÉRICAS E, II SIMPÒSIO EUROPA-AMERICA, 3., 2006, Medellin, Proceedings, Medellin/Colombia. **Proceedings...** Colombia, p. 18-23, 2006.

BEZERRA, J. E. F.; SILVA JUNIOR, J. F. da; LEDERMAN, I. E. **Pitanga (*Eugenia uniflora* L.). Jaboticabal**: FUNEP, p. 30, 2000.

BOBBIO, F. O.; BOBBIO, F. O. Natural pigments. In: INTRODUÇÂO À QUiMICA DE ALIMENTOS. 2 ed., 1 reimpr Sao Paulo: Livraria Varela, cap. VI p. 191-22, 1992.

BRAZIL. Ministry of Agriculture, Supply and Agrarian Reform. Resolution No. 5, November 13, 2000. Identity and quality standards (PIQ) for fermented milks. **Diàrio Oficial [da] Repùblica Federativa do Brasil**, Brasilia, DF, 27 nov. 2000. Section 1, p. 9-12.

BRAZIL. Normative Instruction No. 1, of January 7, 2000. General Technical Regulation for setting Identity and Quality Standards for Fruit Pulp. **Diàrio Oficial [da] Repùblica Federativa do** Brasil, Brasilia, DF, 10 jan. 2000, Seçao 1, p. 54.

BRAZIL. Ministry of Agriculture, Livestock and Supply. Department for the Inspection of Products of Animal Origin. Normative Instruction No. 51, of September 18, 2002. Approves and formalizes the technical regulation on the identity and quality of refrigerated raw milk. **Official Gazette**, n. 172, 2002. Section I, p.13-22.

BRAZIL. Technical Regulations for the Production, Identity and Quality of Fermented Milks. Normative Instruction No. 46. **Official Gazette**, Brasilia: Ministry of Agriculture, October 23, 2007. Section I, p. 4.

BRAZIL. Obtaining quality milk and making dairy products. Document 154. ISSN 1806 - 9193, June, 2006. p. 24.

BRAZIL. Normative Instruction No. 62, of December 29, 2011. Technical regulations for the production, identity, quality, collection and transportation of milk. Brasilia, DF, 2011. p. 24.

BRITO, M. A. V. P.; BRITO J. R. F.; PORTUGAL J. A. B. Identification of bacterial contaminants in raw milk from refrigeration tanks**. Revista do Instituto de Laticinios Càndido Tostes**. v. 57, p. 47-52, 2002.

BROINIZI, P. R. B. et al. Evaluation of the antioxidant activity of phenolic compounds naturally present in by-products of cashew pseudofruit (*Anacardiumoccidentale L.*). **Ciência e Tecnologia de Alimentos**, Campinas, v. 27, n. 4, p. 890-896, 2007.

BOLINI, H. M. A.; MORAES, P. Thesis shows that sensory analysis would increase yogurt production. **Jornal da Unicamp**, v. 11, p. 253, 2004.

BORGES, K. C. de F. et al. Pulp yield and fruit and seed morphometry of pitangueira-do-cerrado. **Revista Brasileira de Fruticultura,** Jaboticabal - SP, v. 32, n. 2, p. 471-478, June, 2010.

BORGES, K. C.; MEDEIROS, A. C. L., CORREIA, R. T. P. Cajà (*Spondias lùtea* L.) flavored buffalo milk yogurt: Physico-chemical characterization and sensory acceptance among 11 to 16 year olds. **Alimentos e Nutriçâo, Araraquara**, v. 20, n. 2, p. 295-300, Apr/Jun, 2009.

BUAINAIN, A. M.; BATALHA, M. O. **Fruit production chain**. Brasilia: IICA/MAPA/SPA, v.7, p. 102, 2007.

BUZI, K. A.; PINTO, J. P. A. N.; RAMOS, P. R. R.; BIONDI, G. F. Microbiological analysis and electrophoretic characterization of mozzarella cheese made from buffalo milk. **Ciência e Tecnologia de Alimentos**, Campinas, v. 29, n. 1, p. 711, 2009.

CAMLOFSKI, A. M. O. **Characterization of the fruit of the cherry tree (Eugenia involucrata DC) with a view to its technological use**. 2008. 107 f. Dissertation (Master's in Food Science and Technology) - State University of Ponta Grossa, Ponta Grossa, 2008.

CHITARRA, M. I. F.; CHITARRA, A. B. **Post-harvest of fruit and vegetables: physiology and handling**. 2. ed. Lavras: UFLA, p. 783, 2005.

COELHO, K. O. et al. Detection of the physicochemical profile of buffalo milk samples using automated analyzers. **Ciência Animal Brasileira**, v. 5, n. 3, p. 167-170, jul./set. 2004.

COSTA G. N. S. et al. Development of a Juçai Flavored Yogurt (*Euterpe edulis Martius*): Physico-chemical and Sensory Evaluation. **Revista Eletrônica TECCEN, Vassouras**, v. 5, n. 2 p. 43-58, mai./ago., 2012.

COULTATE, T. P. **Food:** The chemistry of its components. 3. ed. Porto Alegre: Artmed, p. 368, 2004.

CUNHA NETO, O. C. et al. Evaluation of natural yogurt produced with buffalo milk containing different levels of fat. **Ciência e Tecnologia de Alimentos, Campinas**, v. 3, n. 25, p.448-453, jul/set., 2005.

CUNHA NETO, O. C.; OLIVEIRA, C. A. F. Aspectos da qualidade microbiológica do leite de bùfala. **Food Hygiene**, v. 17, n. 110, p. 18-23, 2003.

DAMODARAN, S.; PARKIN, K. L.; FENNEMA, O. R. **Fennema's Food**

Chemistry. 4. ed., Porto Alegre: Artmed, p. 900, 2010.

DALMASSO, A.; CIVERA, T.; LA NEVE, F.; BOTTERO, M. T. Simultaneous detection of cow and buffalo milk in mozzarella cheese by Real-Time PCR assay. **Food Chemistry**, London, v. 124, n. 1, p. 362-366, 2011.

DOWNHAM A, C. P. Colouringourfoods in thelastandnextmillennium. **International Journal of Food Science Technology.** v. 35, n. 1, p. 5-22, 2000.

FARIA, C. P.; BENEDET, H. D.; LE GUERROUE, J. Analysis of buffalo milk fermented with *Lactobacillus casei* and supplemented with *Bifidobacteriumlongum*. **Semina: Ciências Agrârias**, Londrina, v. 27, n. 3, p. 407-414, jul./set., 2006.

FAO. FAOSTAT: production-crops. Available at:

<http://faostat.fao.org/site/567/DesktopDefault. aspx?PageID=567#ancor>. Accessed on: June 6, 2011.

FAVARO-TRINDADE, C. S. et al.

Sensoryacceptabilityandstabilityofprobioticmicroorganismsandvitamin Cin

fermented acerola (*Malpighiaemarginata* DC.) ice cream. **Journaloffoodscience**, v. 71, n. 6, p. 492-495, 2006.

FENNEMA, O. R. Quimica de los alimentes. 2. ed. Zaragoza: Acribia, p. 1258, 2000.

FELLOWS, P. J. **Food Processing Technology:** principles and practices. Sao Paulo: Artmed, 2006.

FRANCO, G. **Tabela de composiçâo quimica de alimentos**. 9. ed. Rio de Janeiro: Atheneu, p. 307, 2006.

FREITAS, C. A. S. et al. Acerola: Production, composition, nutritional aspects and products. **Revista Brasileira de Agrociência**, Pelotas, v. 12, n. 4, p. 395-400, 2006.

FURTADO, M. M. Composiçao centesimal do leite de bùfala na zona da mata mineira. **Revista do Instituto de Laticinios "Càndido Tostes"**, v. 35, n. 211, p.4347, 1980.

FOOD AND AGRICULTURE ORGANIZATION - FAO. FAO **statistical yearbook 2013.** Rome: FAO, 2013. p. 289, 2013.

GUERRA, R. B.; NEVES, E. C. A.; PENA, R. S. Caracterizaçao e processamento de leite bubalino em pó em secador por nebulizaçao. **Ciência e Tecnologia de Alimentos**, Campinas, v. 25, n. 3, p. 443-447, 2005.

GEMTCHÜJNICOV, I. D. **Manual of plant taxonomy: plants of economic, agricultural, ornamental and medicinal interest**. Sao Paulo: Ceres, p. 368, 1976.

KOH, E.; MITCHELL, A. E. **Trends in the Analysis of Phytochemicals - Flavonoids and carotenoids**. In: MESKIN, S.M.; BIDLACK, W.R.; RANDOLPH, R. K. Phytochemicals - Aging and Health. Buena Park: CRC Press. p. 39-76. 2006.

KROLOW, A. C. R. **Coffee-flavored whole milk yogurt.** Technical note 193. Online version. ISSN 1806-9185. Pelotas, RS. Dec. 2008.

KUSKOSKI, E. M. et al. Wild tropical fruits and frozen fruit pulps: antioxidant activity, polyphenols and anthocyanins. **Ciência Rural**, Santa Maria, v.36, n. 4, p. 1283-1287, 2006.

KUSKOSKI, E. M. et al. Antioxidant activity of anthocyanin pigments. **Ciência e Tecnologia de Alimentos,** v. 24, n. 4, p. 691, 2004.

LIMA, V. L. A. G. de; MÉLO, E. A.; LIMA, D. E. S. Total phenolics and carotenoids in pitanga. **Scientia agricola**, Piracicaba, v. 59, n. 3, p. 447-450, 2002.

LIMA, V. L. A. G. et al. Physico-chemical and sensory characterization of purple pitangas. **Revista Brasileira de Fruticultura**, Jaboticabal, v. 22, n. 3, p. 382-385, 2000.

LOPES, A. L.; MATTIETTO, R. de A.; MENEZES, H. C. de. Stability of pitanga pulp under freezing. **Ciência e Tecnologia de Alimentos**, v. 25, n.3, p. 553559, 2005.

LOPES, F. A. **Characterization of the Productivity and Quality of Buffalo Milk in the Southern Mata Zone of Pernambuco**. 2009. 48 f. Dissertation (master's degree) - Federal University of Pernambuco, Recife, PE, p. 45, 2009.

LOURENÇO JUNIOR, J. B.; GARCIA, A. R. Overview of bubalinoculture in the Amazon. In: ENCONTRO INTERNACIONAL DA PECUÂRIA DA AMAZÔNIA,

1., 2008, Belém, **Anais...** Belém: FAEPA; Instituto Frutal; SEBRAE-PA, 2008.

MARTÌNEZ-FLÓREZ S. et al. Flavonoids: properties and antioxidant actions. **Nutrición Hospitalaria,** n. 17, v. 6, p. 271-8, 2002.

MACEDO, M. P. et al. Physico-chemical composition and milk production of Mediterranean buffaloes in the west of the state of Sao Paulo. **Revista Brasileira de Zootecnia**, v. 30, n. 3, 2001.

MAZZA, G. J. Anthocyaninsandhearthealth. **Annalidell'IstitutoSuperiorediSanità**. v. 43. p. 369-374. 2007.

MATHIAS, T. R. S.; ANDRADE, K. C. S.; ROSA, C. L. S.; SILVA, B. A. Evaluation of the rheological behavior of different commercial yogurts. **Brazilian Journal of Food Technology,** Campinas, v. 16, n. 1, p. 12-20, 2013.

MEDEIROS, F. C. et al. Centesimal composition of yogurts sold in the municipality of Bananeiras-PB. **II Jornada Nacional da Agroindùstria, Bananeiras**, Dec. 2007.

MÉLO, E. A.; LIMA, V. L. A. G.; NASCIMENTO, P. P. Temperature in the storage of pitanga. **Scientia Agricola**, v. 57, n. 4, p. 629-634, Oct./Dec. 2000.

MINISTRY OF AGRICULTURE, LIVESTOCK AND SUPPLY - MAP. **Cattle and buffaloes**. 2016. Available at: <http://www.agricultura.gov.br/animal/especies/bovinos-e-bubalinos>. Accessed on: 01 Nov. 2018.

MORAES, P. C. B. T. **Evaluation of commercial strawberry-flavored liquid yogurts: consumer study and sensory profile**. 2004. 128 f. Dissertation (Master's Degree) - State University of Campinas, Campinas, SP, p. 128, 2004.

NADER FILHO, A. et al. Estudo da variaçao do ponto crioscòpico do leite de bùfala. **Revista do Instituto** de **Laticinios Càndido Tostes**, Juiz de Fora, v. 39, n. 234, 1984.

NERES, L. S.; LOURENÇO JUNIOR, J. B.; PACHECO, E. A.; MONTEIRO, R. C. R.; SATO, S. T. A.; LIMA, S. C. G.; GARCIA, A. R.; NAHUM, B. S. Buffalo milk yogurt flavored with mango (Mangifera indica L.): sensory acceptance and production cost. **Agroecossistemas**, v. 4, n. 2, p. 79-84, 2012.

NEVES, N. R. P. **Preparation of açai (*Euterpe Oleracea* Mart.) sundae yogurt.** 2011. 62 f. Course Conclusion Work (Graduation in Agroindustrial Technology with emphasis on Food) - Parà State University, Redençao, 2007.

NIJVELDT, R. J. etal . Flavonoids: a reviewofprobablemechanismsofactionandpotentialapplications. **The American Journal of Clinical Nutrition,** v. 74, n. 4, p. 418-25, 2001.

NOBRE, L. N. N.; BRESSAN, J.; SOBRINHO, P. S. C.; COSTA, N. M. B.; MININ, V. P. R.; CECON, P. R. Volume of light yogurt and subjective appetite sensations in eutrophic and overweight men. **Revista Nutriçâo**, Campinas, v. 19, n. 5, p. 591-600, 2006.

VERRUMA-BERNARDI, M. R.; BRANCO, N. C.M.; MAROTE, D. M.J.; DELIZA, R.; ARAÙJO, K. G. de L.; KAJISHIMA, S. **Sensory profile and preference of buffalo milk yogurt. Boletim do Centro de Pesquisa e Processamento de Alimentos**, Curitiba, v. 24, n. 2, p. 443-456, jul/dez., 2006.

VERRUMA, M. R.; SALGADO J. M. Chemical analysis of buffalo milk compared to cow's milk. **Scientia Agricola**, v. 51, p 131-137, 1994.

VIEIRA, R. G.; FALCÂO FILHO, R. S.; DUARTE, T. F.; PESSOA, T. R. B.; QUEIROGA, R. C. R. E.; MOREIRA, R. T. Acceptability and sensory preference of cheese curds elaborated with female buffalo, goat and cow milk. **Revista do Instituto da Laticinios Càndido Tostes**, Juiz de Fora, v. 363, n. 63, p. 12-16, 2011.

OLIVEIRA, F. M. N.; FIGUÊIREDO, R. M. F.; QUEIROZ, A. J. M. Comparative analysis of whole, formulated and powdered pitanga pulps. **Revista Brasileira de Produtos Agroindustriais**, Campina Grande, v.8, n.1, p. 25-33, 2006.

OLIVEIRA, R. L. Chemical composition and fatty acid profile of milk and mozzarella from buffaloes fed different sources of lipids. **Arquivo Brasileiro de Medicina, Veterinària e Zootecnia**, v.61, n.3, p.736-744, 2009.

OLIVEIRA, K. A. M. et al. Development of araticum yogurt formulation and study of sensory acceptance. **Alimentos e Nutriçâo**, Araraquara v.19, n.3, p. 277281, jul./set.,

2008.

OLIVEIRA, F. M. N. Drying and storing pitanga pulp. 2006. 197 f. Dissertation (Master's in Agricultural Engineering) - Federal University of Campina Grande, Paraiba, April, 2006.

ORDÓNEZ, J. A. **Food Technology: Food Components and Processes**. Porto Alegre: Artmed, 2005.

OZELA, E. F. **Characterization of flavonoids and pigment stability of bertalha fruits (*Basella rubra*, L.).** 2004. 88 f. Thesis (Doctorate in Food Science and Technology) - Federal University of Viçosa, Viçosa, 2004.

PAZMINO-DURAN, E. A. et al. Anthocyanins from Oxalis triangularis as potential food colorants. **Food Chemistry**, London, v. 75, n. 2, p. 211-216, 2001.

PEREIRA, V. R. **Ascorbic acid - characteristics, mechanisms of action and applications in the food industry**. Federal University of Pelotas, Rio Grande do Sul. Department of Food Science Bachelor's Degree in Food Chemistry Food Seminar Course, 2008.

Petrifilm 3M: **Plates for counting Escherichia coli. Instructions for use. 3M do Brasil Ltda**. Microbiology. St Paul, MN 55144-1000.

PINTO, P. M. Z.; MORAIS, A. M. M. B. Boas Prâticas para a Conservaçà de Produtos Horto fruticolas. Porto: Association for the School of Biotechnology of the Catholic University, 2000.

RAHMAN, M. S.; RUIZ, J. F. V. Food Preservation by Freezing. In: RAHMAN, M. S. **Hand book of Food Preservation**, Boca Raton: CRC Press, p. 635-657, 2007.

RIBEIRO, E. P; SERAVALLI, E. A. G. **Quimica de Alimentos**. 2.ed. Sao Paulo: Edgard Blücher, p. 184, 2007.

RIBEIRO, L. O.; MENDES, M. F.; PEREIRA, C. S. S. Evaluation of the centesimal composition, mineral and anthocyanin content of juçai pulp (*Euterpe edulis Martius*). **Revista TECCEN,**Vassouras, RJ, v. 4, n. 2, p. 5-16, Sep./Dec. 2011.

ROGEZ, H. Açai: **Preparation, composition and improved conservation**. Belém: Editora da Universidade Federal do Parà - EDUFPA, p. 313, 2000.

ROCHA, C. et al. Yogurt made from buffalo milk flavored with Cerrado fruits. **Bulletin of the Food Processing Research Center**, v. 22, n. 1, p.97-106, jan/jun., 2004.

RODAS, M. A. B. et al. Physical-chemical and histological characterization and viability of lactic acid bacteria in yogurt with fruit. **Ciência e Tecnologia de Alimentos**, v. 21, n. 3, Campinas, Sep/Dec, 2001.

RODRIGUES, C. F. C.; IAPICHINI, J. E. C. B.; LISERRE, A. M.; SOUZA, C. B.; FACHINI, C.; REICHERT, R. H. Oportunidades e desafios da bubalinocultura familiar da regiào sudoeste paulista. **Revista Tecnologia & Inovaçâo Agropecuària**, Sao Paulo, v. 1, p. 100-109, 2008.

SARNI-MANCHADO, P. S. et al. Stability and color of unreported wine anthocyanin-derived pigments. **J. Food Sci.**, Chicago, v. 61, n. 5, p. 938-941, 1996.

SANCHOTENE, M. C. C. **Frutiferas nativas ùteis a fauna na arborização urbana**. 2 ed. Porto Alegre: Sagra, p. 304, 1989.

SANTINI, G. A.; BERNARDES, O. ; SCARPELLI, J. U. Analysis of the commercial relations of the processing segment of milk and buffalo milk derivatives in the state of Sao Paulo. **Informaçôes Econômicas**, v. 43, n. 5, p. 69-84, 2013.

SANTOS, M. S. et al. Physico-chemical characterization, extraction and analysis of pectins from Campomanesia Xanthocarpa B. (Gabiroba) fruits. **Semina: Ciências Agrârias**, Londrina, v. 30, n. 1, p. 101-106, 2009.

SALGUEIRO, F. et al. Evenpopulationdifferentiation for maternal andbiparental gene markers in Eugenia uniflora, a widely distributed species from the Brazilian coast al Atlanticra in forest. **Diversity and Distributions**,Oxford, v. 10, n.3, p. 201-210, 2004.

SÂO PAULO. Secretariat of Agriculture and Supply of the State of São Paulo. SAA Resolution No. 24 of August 1, 1994. Technical standards for the production and classification of products of animal origin, inspection activities and inspection of products of animal origin.

SILVA, S. L. A.; FERREIRA, G. A. L.; SILVA, R. R. The search for vitamin C. **Quimica Nova na Escola**, n. 2, nov. 1995.

STATSOFT, Inc. STATISTICA (**data analysis software system**), version 7. 2004.

SOUZA R. M. S. R. **Buffalo milk yogurt with Amazonian fruit pulp:** quality parameters. Universidade federal fluminense-DFF, Programa de pós-graduaçao em medicina veterinària doutororado interinstitucional, Niterói, Rio de Janeiro, 2011.

TACO, **Tabela Brasileira de Composiçâo de Alimentos** - NEPA/UNICAMP - Versao II - 2. ed. Campinas, SP: NEPA/UNICAMP, p. 113, 2006.

TEIXEIRA, G. H. A. **Fruits of the bacuriz tree (Platoniainsignis Mart): characterization, quality and conservation**. 2000. 106 f. Dissertation (Master's Degree in Agronomy) - Faculty of Agricultural and Veterinary Sciences, São Paulo State University, 2000.

SANTOS, A. F. dos; SILVA, S. DE M.; MENDONÇA, R. M. N.; SILVA, M. S. DA; ALVES, R. E.; ALMEIDA, H. Physical changes during the ripening of red and purple pitanga (*Eugenia uniflora* L.). **Anais...** Tegucigalpa, Honduras, 2002.

TEIXEIRA NETO, F. **Nutriçâo clinica Rio de Janeiro:** Guanabara Koogan. p. 76, p. 128-129 e 288, 2009.

VALLILO, M. I. et al. Chemical composition of the fruits of *Campomanesiaadamantium (Cambessédes)* O. BERG. **Ciência e Tecnologia Alimentos**, Campinas, v. 26, n. 4, p.805-810, 2006.

VANNUCCHI, H.; ROCHA, M. de M. Fully recognized functions of nutrients: Ascorbic acid (Vitamin C). **ILSI Brasil International Life Sciences Institute do Brasil**, v. 21, p. 3-11, 2012.

VARNAN, A. H.; SUTHERLAND, J. P. **Leche y produtos làcteos: tecnologia, quimica y microbiologia**. Zaragoza (Spain): Acribia, p. 476, 1995.

WANG, H.; NAIR, M.G.; STRASBURG, G. M; CHANG, Y. C.; BOOREN, A.M.;

GRAY, J. I. Antioxidant and anti-inflammatory activities of anthocyanins and their aglycon, cyanidin, from tart cherries. **Journal of Natural Products.** v. 62, n. 2, p.294-6. 1999.

WILKINSON, J. **Patterns of competition and regulation in world industry.** In: ESTUDO DA COMPETITIVIDADE DA INDÙSTRIA BRASILEIRA: O COMPLEXO AGROINDUSTRIAL [online]. Rio de Janeiro: Edelstein Center for Social Research, p. 56-69, 2008. ISBN 978-85-99662-64-9.

Printed by Books on Demand GmbH, Norderstedt / Germany